BUILDING ON A DREAM

THE SYDNEY OPERA HOUSE

Nicole K. Orr

PURPLE TOAD PUBLISHING

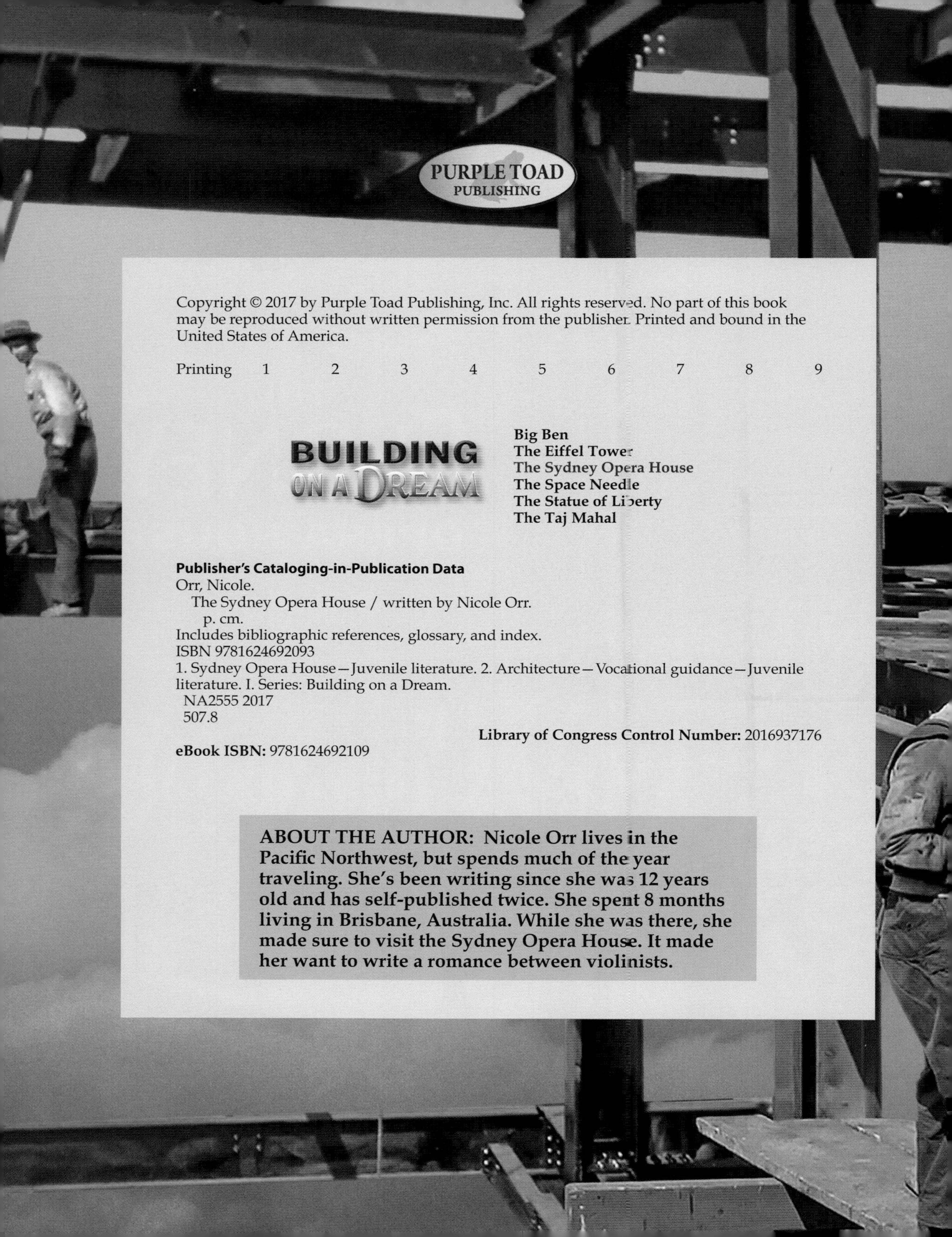

Printing 1 2 3 4 5 6 7 8 9

BUILDING ON A DREAM

Big Ben
The Eiffel Tower
The Sydney Opera House
The Space Needle
The Statue of Liberty
The Taj Mahal

Publisher's Cataloging-in-Publication Data
Orr, Nicole.
The Sydney Opera House / written by Nicole Orr.
p. cm.
Includes bibliographic references, glossary, and index.
ISBN 9781624692093
1. Sydney Opera House—Juvenile literature. 2. Architecture—Vocational guidance—Juvenile literature. I. Series: Building on a Dream.
NA2555 2017
507.8

Library of Congress Control Number: 2016937176

eBook ISBN: 9781624692109

ABOUT THE AUTHOR: Nicole Orr lives in the Pacific Northwest, but spends much of the year traveling. She's been writing since she was 12 years old and has self-published twice. She spent 8 months living in Brisbane, Australia. While she was there, she made sure to visit the Sydney Opera House. It made her want to write a romance between violinists.

CONTENTS

Tourists and locals alike come out to see the sails of the Opera House, especially when ablaze with light.

CHAPTER 1

Lighting the Sails

On a mild winter night outside the Sydney Opera House, people stand on tiptoe to get a better view. They hold up their cell phones to take videos. The crowd gasps, masking the sound of the Sydney River splashing below.

What are they looking at? They are staring at a shower of color. Projected onto the Sydney Opera House is a parade of images. First, there are trees bathed in sunlight. Next are fireworks exploding, then pinball machines—with sound effects: *Ding, ding, ding!*

Welcome to Vivid Sydney. Since 2009, this "festival of creativity" has been the largest of its kind. It combines light, music, and ideas for new technology. For the first seven years, the event lasted 18 nights. In 2016, it lasted for 23 nights. Although the performances, activities, and guests change every year, the event always has three parts: Vivid Music, Vivid Ideas, and Vivid Light.

Drawing music lovers from everywhere, Vivid Music stops at venues all across Sydney, including the Sydney Opera House. From small bands to full orchestras, this event pulls in some big names. Florence and the Machine, the Cure, and Yo Gabba Gabba are a few examples.

Each year, Vivid Ideas presents a lecture series from people who really make a difference. Speakers include television producers, computer designers, writers, comedians, and activists. These "game changers" know how to turn ideas into reality.

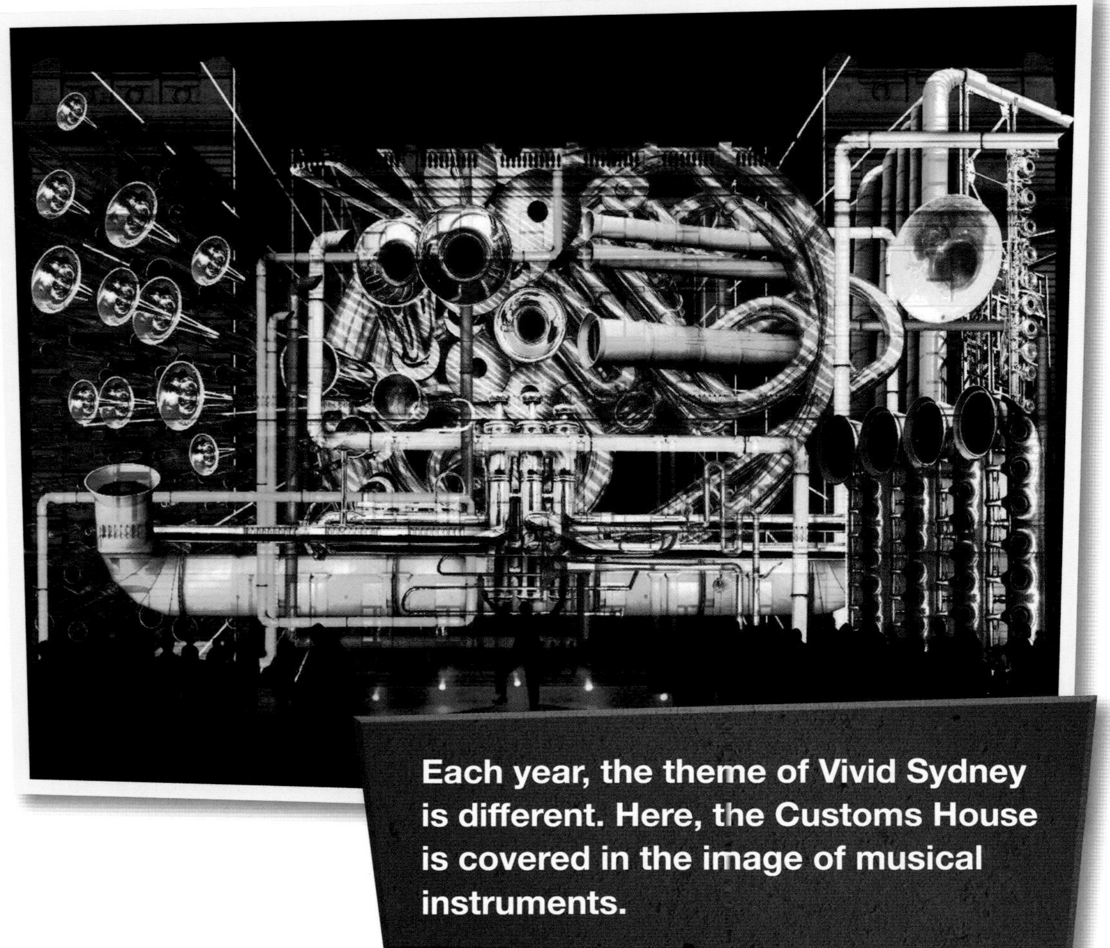

Each year, the theme of Vivid Sydney is different. Here, the Customs House is covered in the image of musical instruments.

Vivid Light creates "light art sculptures." It combines light shows with famous landmarks. Lights flash and shine. Colorful images like carnivals or mountains are shown for all to see.[1]

These gigantic projections mix art with math and engineering. Images that are usually shown on flat surfaces have to bend around corners. They have to make sense on unusual shapes. Doing this takes high-tech computers. It also takes an entire team of scanners and designers. Spinifex Group is the Australian crew that created several Vivid Sydney light shows. In 2013, they used 17 projectors to light up the Opera House. The creative team took many small images and blended their edges together. The result was one huge, seamless picture.[2]

The marvels on display at this magical event would not have been possible without the engineering genius of those who built the Sydney Opera House. The creative talent at Vivid Sydney continues a long tradition of problem solving begun in 1956, when a young architect from halfway around the world won a design contest.

Seen from across the water, the Sydney Opera House is a rainbow of color. More photos are snapped of it on these nights than any other.

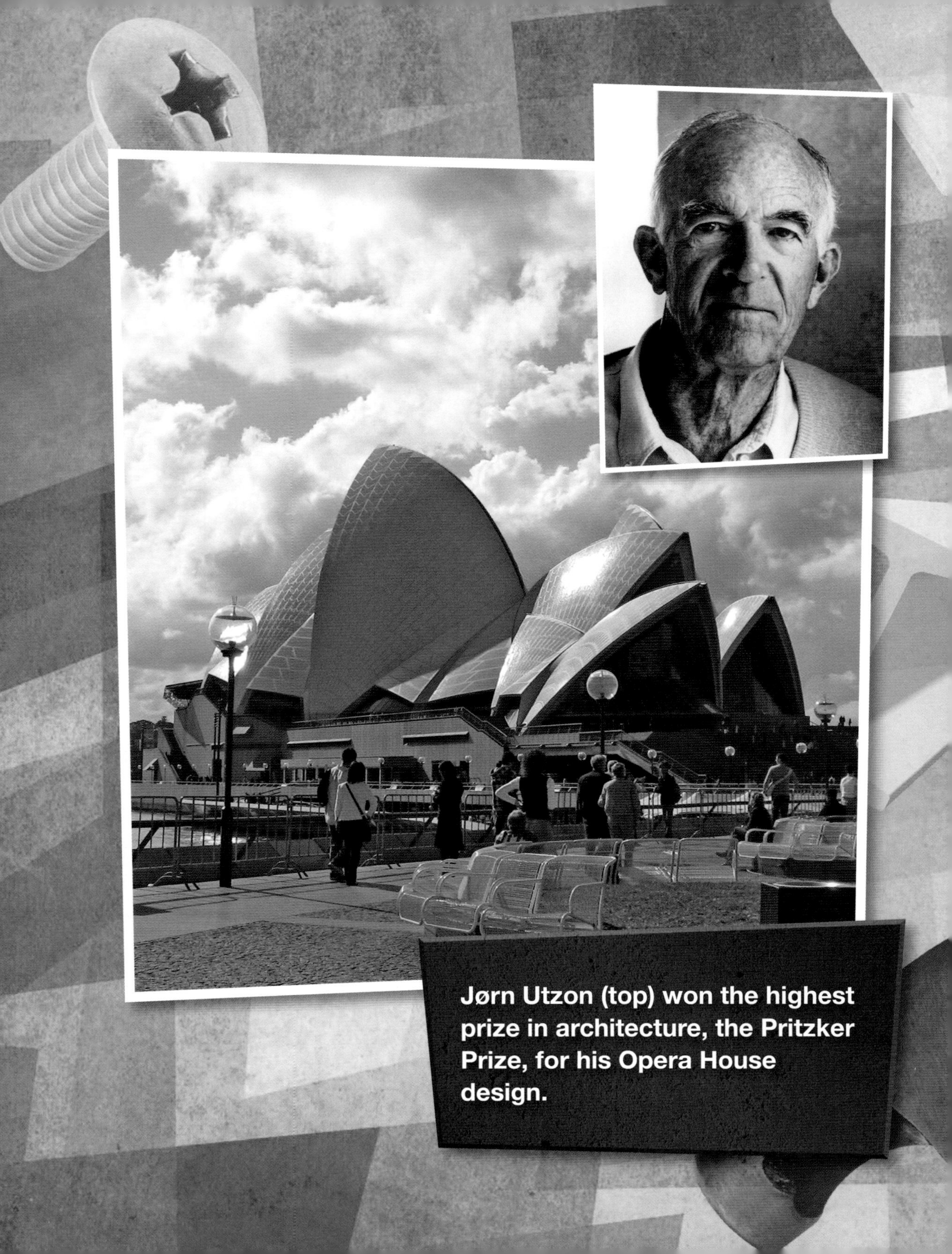

Jørn Utzon (top) won the highest prize in architecture, the Pritzker Prize, for his Opera House design.

CHAPTER 2

It Began with a Dream

The Sydney Opera House is the most recognizable building in Australia. Built on the tip of Bennelong Point in Sydney Harbor, it looks like a sailboat under full sail. This sculptural building is in scale with other large structures around the harbor, including a massive sandstone cliff face and the Harbour Bridge. Viewed from any side or angle, the Sydney Opera House is magnificent.

The exterior of the building is not its only marvel. The structure has two main halls—one for concerts and the other for operas. It also has five other performance spaces, restaurants, and meeting rooms.

Jørn Utzon designed the Opera House, even though he had never been to Australia. Raised in Denmark, Utzon spent most of his childhood on the shipyard docks where his father worked. Young Utzon nearly followed in his father's footsteps. He could have easily designed boats instead of buildings.

After getting his diploma from the Royal Danish Academy of Fine Arts, Utzon went out into the world. In Morocco, he saw Islamic architecture and realized something he'd been thinking about for years. He saw the relationship there could be between manmade architecture and its natural surroundings. Though it would be years before he would find a name for it, this was one of his first steps on the way to additive architecture. To Utzon, it would mean creating designs based on the growth patterns found in nature.[1]

Utzon continued to travel, experiencing new cultures and adding to them. He designed buildings in Sweden and other parts of Europe. He participated in competitions. He and his wife, Lis, whom he had met at college, toured the United States and Mexico. They drove through the countryside, meeting famous architects such as Frank Lloyd Wright and Richard Neutra. The trip inspired Utzon. He was ready to create something bigger. He was ready to surprise the world.

A House and a Home

Utzon wanted to start a family. He also wanted to build a home for that family. In his hometown of Hellebaek, Utzon drew up the designs for the house he had in mind. He wanted one side of his house to have no

Utzon's house was so unusual, he did most of his planning along the way. His team of builders had to do their jobs without written plans.

Utzon was given a very small area on which to build. He had to be very careful about where he built it.

windows or doors. He wanted the other side to be entirely glass. In order for this to work, he had to take into account the sun, the ground, and even the forest.

Other architects were intrigued by Utzon's ideas. They visited the work in progress. One of them was Richard Neutra. After the house was completed in 1952, Neutra took two pictures from inside the house. He said that if he took these photographs home to California, his colleagues would think he had taken them in two different buildings. This was because the different parts of Utzon's house were so different from each other.

Utzon's house in Hellebaek was the first one to showcase his architectural ideas. It grabbed the attention of people around the world. However, it wasn't until he submitted five sketches to an Australian competition that he truly gained the spotlight. While Utzon had always had big ideas, he had never before imagined something as big as the Sydney Opera House.

Design sketches by Jørn Utzon

Eero Saarinen built furniture for his father. Luckily for Utzon, he was also skilled in architecture.

In 1956, a call went out that there was an international competition for designing an opera house for Sydney, Australia. It would not be an easy project. The design had to include two performance halls. One would be used for opera singers. The other would be used for symphony concerts. That wasn't all! The opera house also had to have a restaurant and two meeting rooms.[1]

A year after the competition was announced, submissions were closed. Two hundred and thirty-three designs had been sent in, from 32 countries. The decision would take 10 days.

One of the judges, Eero Saarinen, arrived late. He almost missed his chance to make one of the most important decisions of his life. If it weren't for him, Utzon's designs might have stayed where they were—in the rejection pile.

Saarinen saw Utzon's ideas and quickly placed them back on the table. The world might have ended up with quite a different Opera House if not for this last-minute decision.

Breaking Ground

The process of building the Opera House was divided into three stages. In Stage I, the foundation would be dug and the podium built. Columns to hold up the roof would be installed. Stage II would see the roof go up. Finally, Stage III would complete the interior.

March 2, 1959, was a breezy day in Sydney, Australia. Jørn Utzon and Joe Cahill wore heavy suit jackets to ward off the wind. Cahill, the

The site for the Opera House was Bennelong Point. "Bennelong" was the name of one of the first Aboriginal men to learn English. The point is named after him.

prime minister of New South Wales, had approved the building of the Opera House. He bent over a large square stone. The stone had been laid where the axis of the two main halls of the Opera House would intersect. In one hand he had a hand-made screwdriver. In the other he had a special pin. Cahill used the screwdriver and pin to attach a plaque to the stone.

Music was played and speeches were given. As the excitement swelled, the two men shared a grin. This ceremony began Stage I of construction.

Stage I (1959–1963)

From the start, construction was riddled with problems. By the time the first stage started on March 2, the project was already 47 weeks behind schedule. Heavy storms and last-minute changes to contracts delayed the start, but not as much as waiting for the finished plans.

Despite these issues, the building team pushed on. Because of everyone's haste, the podium columns ended up being too weak to support the roof. They had to be knocked down and built again, causing further delays. By the time Stage I was completed, the project was two years behind.

After an old railway shed was demolished, digging began for the podium. It would serve as the foundation for the opera house. Actors and musicians would prepare for their shows in its dozens of rooms.

Stage II (1963–1967)

For three years, Utzon tangled with the problem of how to build the roof. He examined the theaters and temples of ancient Greece. He looked at Gothic churches and old castles. He made many models, searching for the right shape for the roof's shells. He knew that once he found the right design, making the actual pieces would be easy.

Using computers to design a building was a new idea at the time. The Sydney Opera House builders were happy to give them a try. The project team created virtual models. They used their computers to test their theories for the shells.

Ove Arup questioned whether their designs would be able to hold up under high pressures. The team made scale models and sent them to Southampton University, a school that had wind tunnels. The models were put inside the wind tunnels to see how the building handled strong winds. Arup's concerns were proven correct. The models failed. He and Utzon would have to find a new solution.[2]

The first model of the Opera House was a failure. The site for the Opera House would be very windy.

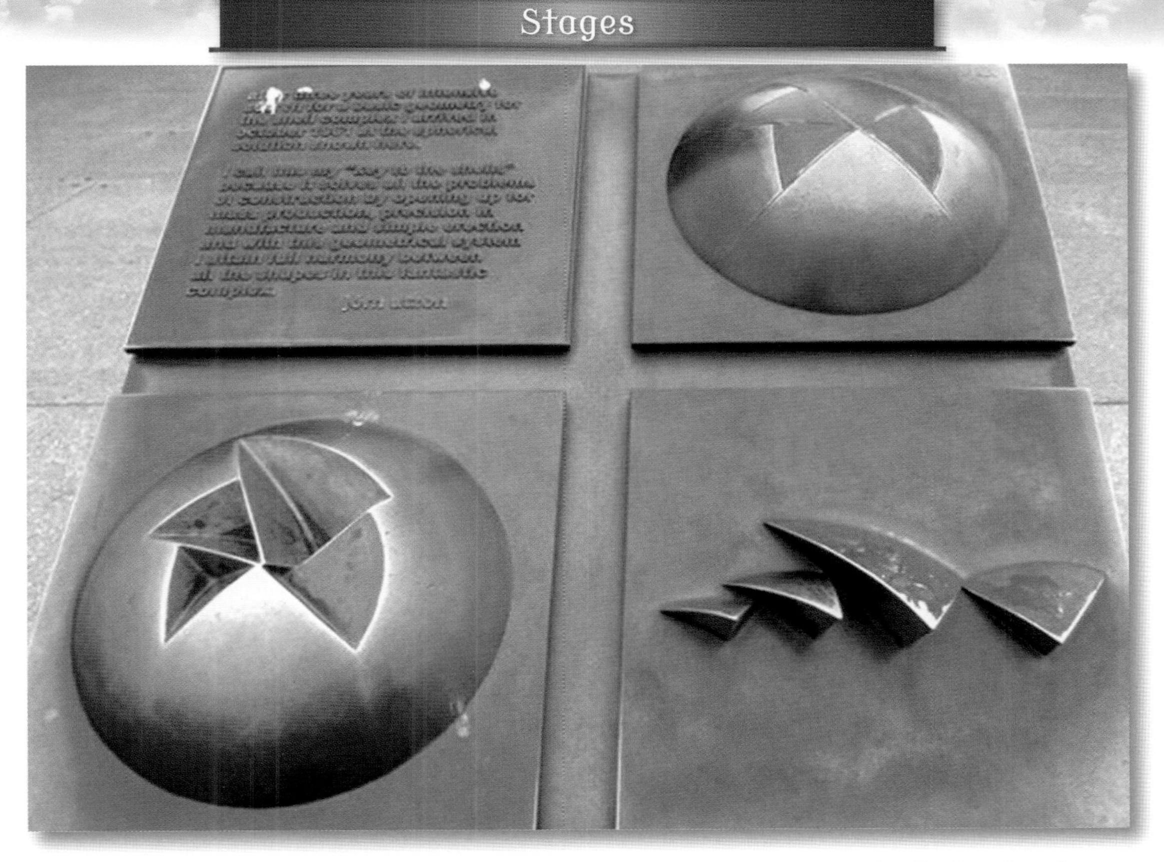

A plaque outside the Opera House describes Utzon's Spherical Solution.

Just who came up with the solution has been argued for years. The most common theory has been that Utzon, abruptly, approached his creative team. He had one of his colleagues get him an orange. Utzon peeled it and then held out the pieces. "It must be a sphere!" he told them.

Using a sphere to construct the roof of a building was an entirely new concept. This design would come to be known as the Spherical Solution.[3] This time, the models passed the wind tunnel tests.

Utzon worried that once completed, the shells of the Opera House would glare more than shine. How could he fix this? Utzon asked for help from a Swedish stoneware company. They used a mixture of clay

Sydney Tiles

and crushed stones to give the shells the right effect. The colors were a glossy white and cream. The Opera House made this combination famous. This special concrete became known as Sydney Tile.

Stage III (1967–1973)

Just as Stage III was getting started, the project ended for Utzon. Davis Hughes, the new Minister of Public Works, stopped the funding. Utzon sent in complaints, along with receipts that proved what he was owed. Hughes thought these expenses were ridiculous.[4] In 1966, Utzon could no longer pay for materials or workers. He resigned.

Utzon and his family left Australia and returned to Denmark. A man named Peter Hall was brought in to replace him. Rumors quickly spread about the Opera House. One was that Utzon had taken too much time and spent too much money. While $22 million was spent in the nine years Utzon was in charge, much more was spent after he left. Peter Hall's team spent nearly $80 million over the following seven years.

Locals in Sydney were so upset when Utzon left, they staged a protest. It did not work.

Queen Elizabeth II, the Duke of Edinburgh, and Ben Blakeney, a descendant of Bennelong, were all present at the opening of the Opera House.

Many experts believe that if Utzon had stayed until the Opera House was finished, the overall expense would have been much lower than the whopping $102 million it ended up costing.

Showing the Opera House to the World

On October 20, 1973, the Sydney Opera House was finished. Queen Elizabeth II was there for the grand opening. It felt a lot like that windy day had when Utzon and Cahill had attached the plaque in 1959. There was music again. There were speeches. The only difference was that neither Utzon nor Cahill was there to see it.

The people of Australia were upset at the loss of Utzon. On March 3, 1966, three thousand people signed a petition demanding that Utzon be rehired. It didn't work. More than three decades passed before the Sydney Opera House Trust reached out to Utzon.

Chapter 3

The area of town where the Opera House sits is called Circular Quay.

When the Trust asked him to be a design consultant, Utzon was honored. He wasn't the kind of person to hold a grudge, not even against Davis Hughes. Though he worked from home and did not go back to Sydney, he gladly wrote a list of guidelines. These described how to maintain and possibly expand the Opera House.

The orchestra pit manages to serve the musicians, despite its size.

Utzon's designs were very specific. When he was removed from the project, changes were made that caused many problems. For example, Peter Hall switched the concert and opera halls. Now people would be singing in the concert hall and playing music in the opera hall. One thousand seats were added to the concert hall, making it extremely cramped.

The opera hall's pit, where the orchestra sets up, is so tiny that the musicians cannot play comfortably. There is hardly enough room for their elbows, let alone their violin bows. More than that, the space is so small, they cannot hear the different parts of the orchestra. This makes it almost impossible to play together.[5]

When Robeson finished singing at the Opera House, the building crew gave him a hard hat with his name on it.

Taking a Tour

Picture a November afternoon. Construction of the Opera House has barely begun. About 250 workers have just stopped to have lunch. Wearing hardhats, the workers sit on scaffolding. Can you imagine their surprise when Paul Robeson walks up and starts singing? A one-time football player, Robeson sang two songs, "Ol' Man River" and "Joe Hill." He sang them a cappella. The workers hummed along with him. After the performance, they asked Robeson to autograph their gloves.[1]

The Sydney Opera House has come a long way since Robeson performed the Opera House's first concert. In 2007, the United Nations named the Sydney Opera House a World Heritage Site. It has also been added to the State Heritage Register.

The Opera House is 606 feet long and 393 feet wide. This means it casts quite a large shadow. The highest point of the building is 219 feet above sea level. That's as high as a 20-story structure. The roof is made of 2,194 precast concrete sections. They weigh up to 15 tons each. Holding them together is 217 miles of tensioned steel cable. More than one million tiles cover the roof.

More than eight million people visit the Opera House each year.[2] Two thousand of them take guided tours. Others attend any of the 3,000 events staged each year. Performances have an annual audience of two million. While the building is open to the public 363 days a year, staff still work 24 hours a day, 7 days a week, 365 days of the year. Put all of this together and the number of workers, musicians, employees, and tourists that have graced the hallways is too high to count.

The glass for the Opera House came all the way from France.

In the hallways, the lighting is remarkable. Large windows let in natural light, but there are thousands of light bulbs in the building. About 15,500 light bulbs are changed every year.

The Opera House has over 1,000 rooms. There are three restaurants, a café, an espresso bar, and opera and theater bars.

While the large Concert Hall seats 2,679, the Utzon room seats only 210. It is used mostly for small, intimate performances. When the architect was told the room would be named after him, Utzon was deeply pleased. He responded, "It gives me the greatest pleasure and satisfaction. I don't think you can give me more joy as the architect. It supersedes any medal of any kind that I could get and have got."[3]

The Utzon Room is the only room true to Utzon's design.

The Concert Hall is the largest of the seven performance venues. It also holds the Grand Organ.

Another Opera House gem is its Grand Organ. It is the largest mechanical organ in the world. With 10,154 pipes, it took ten years to build. It weighs 37.5 tons (75,000 pounds).[4]

The Sydney Opera House is the only opera house to have an actual opera, *The Eighth Wonder*, written about it. Many stories take place there, and couples have gotten married inside.

It has been 60 years since Utzon entered that contest and won the right to build the Opera House. In that time, this building has become the symbol of Australia.

1918 Jørn Utzon is born in Copenhagen, Denmark.

1956 Australia announces a competition to design an opera house.

1957 Utzon wins the contest.

1959 Construction officially begins on the Opera House.

1960 Paul Robeson sings for the construction workers.

1966 Utzon resigns from the project. Peter Hall takes over the project.

1973 The Sydney Opera House officially opens on October 20.

1999 The Sydney Opera House Trust rehires Utzon as a consultant.

2003 Utzon is awarded the prestigious Pritzker Architecture Prize.

2007 The Opera House is named a World Heritage Site.

2008 Utzon dies on November 29.

2009 The first Vivid Sydney festival is held.

2013 The Opera House celebrates its 40th anniversary.

2016 Vivid Sydney honors indigenous artists and culture.

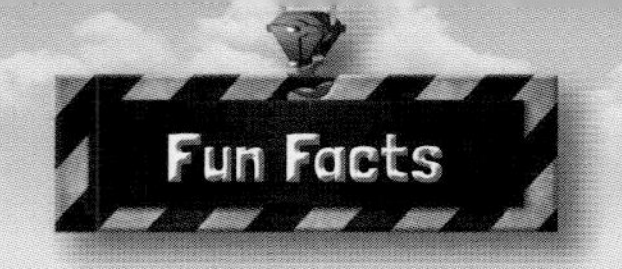

- 10,000 builders were hired to work on the Opera House when construction began in 1959.
- It hosts up to 3,000 events each year.
- There are 2,679 seats in the concert hall.
- There are 1,000 rooms inside the Opera House.
- It has three restaurants. Guillaume at Bennelong is one of them and is considered one of the finest restaurants in Sydney.
- The Opera House is closed to the public only two days a year: Good Friday and Christmas Day.
- The very first opera ever performed there was on September 28, 1973. It was Sergei Prokofiev's *War and Peace*.
- The concert hall's grand organ has 10,154 pipes. It is the largest mechanical organ in the world.
- Inside are seven performance spaces; the Concert Hall, the Opera Theatre, Playhouse, Drama Theatre, The Studio, the Forecourt, and the Utzon Room.
- In 1997, a man named Alain "Spiderman" Robert climbed the outside walls and roofs of the Opera House—all the way to the top—using only his hands and feet.
- The netting above the orchestra pit was installed because a chicken once got loose during a performance and landed on the head of a musician.

Chapter 1

1. Vivid Sydney.
2. Luke Hopewell, "Lighting the Sails: Behind the Scenes on Vivid Sydney's Most Ambitious Project." May 17, 2013, *Gizmodo*.

Chapter 2

1. Danishnet.com, "Jørn Utzon (1918–2008)" Danishnet.com

Chapter 3

1. Sydney Opera House History.
2. Powerhouse Museum, "Model of Sydney Opera House, 1960," Powerhouse Museum Collection Search 2.53.
3. Eric Ellis, "Bennelong Point, Where Utzon's Genius Bore Fruit," *Spectator*, December 10, 2008.
4. "Even When You Lose, You Win," *The Gate*, March 31, 2014.
5. Marina Kemenev, "Sydney's Opera House: Easy on the Eyes, Not on the Ears," *TIME*, October 9, 2011.

Chapter 4

1. Mahir Ali, "Big Voice of the Left Paul Robeson Resounds to This Day," *The Australian*, November 9, 2010.
2. Sydney Opera House Media Room, http://www.sydneyoperahouse.com/about/media_room.aspx
3. Lizzie Porter, "Sydney Opera House: 40 Fascinating Facts," *Telegraph*, October 24, 2013.
4. Mark Fisher, "Sydney Opera House Concert Hall Grand Organ," Organ Historical Trust of Australia, August 2004.

Works Consulted

Ali, Mahir. "Big Voice of the Left Paul Robeson Resounds to This Day." *The Australian*. November 9, 2010. http://www.theaustralian.com.au/arts/big-voice-of-the-left-paul-robeson-resounds-to-this-day/story-e6frg8n6-1225949630309

Anderson, Flemming Bo. *Guide to Utzon* http://www.utzonphotos.com/guide-to-utzon/projects/jorn-utzons-house/

Danishnet.com, "Jørn Utzon (1918-2008)" Danishnet.com http://www.danishnet.com/design/jorn-utzon-danish-designer/

Dellora, Daryl. *Utzon and the Sydney Opera House*. e-Penguin: October 23, 2013.

Ellis, Eric. "Bennelong Point, Where Utzon's Genius Bore Fruit." *The Spectator*. December 10, 2008. http://www.spectator.co.uk/australia/3078101/bennelong-point-where-utzons-genius-bore-fruit/

"Even When You Lose, You Win" *The Gate*. March 31, 2014. http://thegateworldwide.com/london/2014/03/31/even-when-you-lose-you-win/

Fisher, Mark. "Sydney Opera House Concert Hall Grand Organ." Organ Historical Trust of Australia. August 2004. http://www.ohta.org.au/confs/Sydney/GRANDORGAN.html

Hopewell, Luke. "Lighting the Sails: Behind the Scenes on Vivid Sydney's Most Ambitious Project." *Gizmodo*. May 17, 2013. http://www.gizmodo.com.au/2013/05/lighting-the-sails-behind-the-scenes-on-vivid-sydneys-most-ambitious-project/

Kemenev, Marina. "Sydney's Opera House: Easy on the Eyes, Not on the Ears." *TIME*. October 19, 2011. http://content.time.com/time/world/article/0,8599,2097247,00.html

Opera House Project, The. http://theoperahouseproject.com

Own Our House. http://ownourhouse.com.au/

Porter, Lizzie. "Sydney Opera House: 40 Fascinating Facts." *Telegraph.* October 24, 2013. http://www.telegraph.co.uk/travel/destinations/australiaandpacific/australia/sydney/10317099/Sydney-Opera-House-40-fascinating-facts.html

Powerhouse Museum. "Model of Sydney Opera House, 1960." Powerhouse Museum Collection Search 2.53 http://www.powerhousemuseum.com/collection/database/?irn=12041

Sydney Opera House History. http://www.sydneyoperahouse.com/about/house_history_landing.aspx

Vivid Sydney. http://www.vividsydney.com/

Books

Eileen, Hirsch Rebecca. *Australia.* New York: Scholastic, 2012.

Latham, Donna, and Jen Vaughn. *Bridges and Tunnels: Investigate Feats of Engineering with 25 Projects.* Jackson, TN: Nomad Press, 2012.

Lonely Planet. *Not-for-Parents Australia: Everything You Ever Wanted to Know.* Melbourne, Australia: Lonely Planet Publications, 2012.

Ritchie, Scot. *Look at That Building! A First Book of Structures.* Toronto, Canada: Kids Can Press, 2011.

Rough Guides. *The Rough Guide to Australia.* New York: Penguin Group. 2014.

On the Internet

Kids World Travel Guide—Australia
http://www.kids-world-travel-guide.com/australia-facts.html

Science Kids: Engineering for Kids
http://www.sciencekids.co.nz/engineering.html

Time for Kids—Australia
http://www.timeforkids.com/destination/australia

Aboriginal (ab-or-RIH-jih-nul)—Having to do with the native people of Australia.

a cappella (AH kuh-PEL-uh)—Using the voice only, with no instruments to accompany the singer.

additive architecture (AD-ih-tiv, AR-kih-tek-cher)—Buildings that fit well into a natural setting.

autograph (AW-toh-graf)—To sign one's name.

axis (AK-sis)—An imaginary line that divides a shape into two even parts. The plural of *axis* is *axes* (AK-sees).

colleague (KAH-leeg)—A person with whom one works.

consultant (kun-SUL-tunt)—A person who is hired to give his or her expert opinion.

foundation (fown-DAY-shun)—A stone or concrete structure that supports a building from underneath.

intersect (IN-ter-sekt)—To cross.

orchestra (OR-keh-strah)—A group of musicians playing a wide variety of instruments and usually led by a conductor.

podium (POH-dee-um)—The foundation of a building.

scaffolding (SKAF-ul-ding)—A system of platforms on which crews can stand or sit while working at a height above ground.

sculptural building (SKULP-chur-ul BIL-ding)—A building that looks like a piece of artwork that was carved or formed from stone, clay, or metal.

sculpture (SKULP-chur)—A three-dimensional piece of art created by carving or welding.

spherical (SFEER-ih-kul)—Having the shape of a sphere (or ball).

tensioned steel cable (TEN-shund STEEL KAY-bul)—Thick bundles of metal (steel) that can be pulled tight; tensioned steel cables are used in concrete to make it stronger.

venue (VEN-yoo)—An area or room that is used for performances.

PHOTO CREDITS: Cover—John Hill; p. 1—Stephen Karpiniec; p. 4—Nicki Mannix; p. 6—Roderick Eime; p. 7—Ross Foster; p. 8—Roberta Romano; p. 10—Anne-Sophie Ofrim; p. 11—Seier; p. 12—NPS.gov; p. 16—Rory Hide; p. 20—Alex Proimos; pp. 23, 24—Creative Commons; p. 24—Danny Navarro. All other photos—Public Domain. Every measure has been taken to find all copyright holders of material used in this book. In the event any mistakes or omissions have happened within, attempts to correct them will be made in future editions of the book.

Index